INDIAN

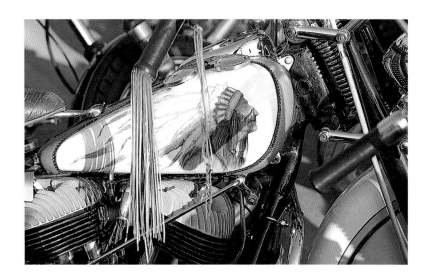

INDIAN

Garry Stuart with
John Carroll

First published in Great Britain in 1994
by Osprey, an imprint of Reed Consumer
Books Limited, Michelin House,
81 Fulham Road, London SW3 6RB and
Auckland, Melbourne, Singapore and Toronto.

ISBN 1 85532 343 5

Project Editor Shaun Barrington
Editor Nick Caldon
Page design Paul Kime/Ward Peacock
Partnership

Produced by Mandarin Offset
Printed and bound in Hong Kong

Half title page
Much of the customising of Indians
revolves around airbrushed variations
on the Indian's head tank design

Title page
The 74 cu in Chief was easily capable
of pulling a sidecar and such a
combination is an impressive vehicle

For a catalogue of all books published by Osprey Automotive
please write to:

**The Marketing Department, Reed Consumer Books,
1st Floor, Michelin House, 81 Fulham Road, London SW3 6RB**

The 1000cc V-twin machines were powerful enough to pull a sidecar comfortably as this beautifully restored outfit shows

Introduction

Think of American made motorcycles and the chances are that the name Harley-Davidson is the one that comes to mind. In truth though they're simply the last surviving US motorcycle manufacturer from a line that has, over the decades, numbered in excess of 250. Their main rival, who went out of business in 1953, was producing successful motorcycles before Harley-Davidson were established. That company? The Hendee Manufacturing Company of Springfield, Massachusetts and their 'Indian Motorcycles'.

Indian was the brand name chosen for the motorcycles and Hendee was the surname of the manufacturer. George M. Hendee was born in 1866 to a family of Spanish origin and became a keen bicyclist to the extent that he won one US National amateur championship in 1886. From here he turned professional and went on to sell bicycles for a living and subsequently to manufacture them. Hendee considered the motorised bicycle to be the logical extension of the pedal bicycle as a means of mass transport. Up until this time the motorcycle was used mainly as a pacing machine in both solo and tandem form, i.e. it was used to slipstream riders going for both endurance and speed cycling records on board cycle tracks around the United States.

One particular pacing machine builder seemed to build more reliable machines than many of his contemporaries; Carl Oscar Hedstrom, a Swedish immigrant to the US. Oscar Hedstrom, as he preferred to be known, was born in 1871, grew up in Brooklyn where he attended school. He became interested in bicycles and eventually became apprenticed to a company that made watch cases. His time had been served by 1892 and Hedstrom worked in a variety of machine shops around New York. His talents soon ensured that the bicycles he made in his spare time found favour with professional racers of the day and he went on to work on pacing machines that were powered by French De Dion internal combustion engines. In 1899 his experience and own theories on the still infant engines enabled him to build a pacing machine that was considerably better than all its predecessors. The Typhoon as it was known was claimed to be the fastest two-wheeled vehicle in the world at that time. It attracted attention from both racers and promoters, including George M. Hendee who was promoting races at The Springfield Coliseum.

Hendee approached Hedstrom with a view to developing a motorised bicycle for commercial production. It appears that Hedstrom had a similar project in mind and he readily agreed and rented workshop space in Middletown, Connecticut. By the spring of 1901 the new machine was ready for a roadtest. Both Hendee and Hedstrom rode the prototype before unveiling it to a large crowd of invited journalists who were, reportedly, duly impressed with many features of the machine and its performance. A spectacular chapter in motorcycle history commences.

Right
Postwar the experimental 841 forks appeared on the Indian Chief as Neil Grieve's motorcycle, seen here with Alan Forbes aboard negotiating a turn, shows

Contents

The Pioneer Years

Hendee and Hedstrom formed the Hendee Manufacturing Company as a partnership. The former was President and General Manager while the latter was Chief Engineer and Designer. They chose the tradename 'Indian' because they felt it typified a wholly American product.

In the summer and autumn of 1901 a handful of motorcycles were produced at Hendee's Springfield bicycle factory. There was no foundry at the bicycle plant and engine castings were manufactured in Illinois by The Aurora Automatic Machine Company (This company briefly produced motorcycles under the tradename of Thor and used the same engine castings). While Hedstrom produced these early models and exhibited a second prototype around the bicycle racing tracks, Hendee raised finance from the Massachusetts business community through the sale of shares in the new company. The pair seem to have been aware of the need to both publicise and promote their products and shipped one of their 1901 models across the Atlantic to an English Show in 1902 The Annual Stanley Bicycle Show.

The company received more orders than it was able to meet but did sell almost 150 machines to members of the public. George M. Hendee and Oscar Hedstrom had proved the worth of both their ideas and their

Jules Frohlich with his 1904 Indian, one of the most original early models known to exist. The Antique Motorcycle Club of America use the machine for reference. Only the seat and tank decal are not original

The 1904 models were slightly improved versions of the 1901 production models and this machine is the 667th to have been made by Hendee and Hedstrom

abilities, now they would open the door to commercial success. Building on the awareness of the need to keep their motorcycles in the public eye they entered machines in a variety of sporting events which rewarded the company with valuable publicity. In 1902 Indian won a ten-mile event on the Brooklyn Cycle Path, the first motorcycle race in the United States. Reporting of this and other Indian successes in cycling magazines and trade publications of the day meant that Indian established themselves as one of the more prominent manufacturers in a fledgling industry. By 1904 sales of Indian motorcycles were paying for the cost of their manufacture rather than there being reliance on the company's capital although further shares were sold in Springfield to raise the wherewithal to pay for additional tooling in order to increase production. In the 1904 sales year the company sold 586 motorcycles and doubled that in 1905 with sales of around 1180. As can be seen from the chapter on single cylinder machines the 1905 models had numerous detail improvements over the earlier ones as a result of Hedstrom's policy of continually updating the product through experimentation.

1911 saw the commencement of work on a huge extension to the Indian factory. Additional land was to allow for further future development. The finance for this was raised through the sale of shares. These were mostly bought by Springfield investors and bankers, something of a mixed blessing. Hendee and Hedstrom felt it necessary to keep the cost of their motorcycles competitive bearing in mind the numerous other motorcycle manufacturers against whom Indians were competing. Chief Engineer Hedstrom wanted to produce machines using only the best quality materials and components whether they were bought in or made by Indian. The upshot of this was that although Indian Motorcycles were selling, in fact demand was increasing, the profit per motorcycle was not great. The sale of successive blocks of shares also meant that the founders had effectively lost control of their company.

During 1912 there were numerous disagreements between the board of directors and Hendee and Hedstrom. The contentious issues were things like the costs of the racing team, expansion of the factory and the profits per motorcycle. The directors were businessmen rather than motorcycling enthusiasts and felt that greater profits were attainable and wanted to maximise their dividends. As is often the case in business, the money men held sway and, though it may not have been known, jeopardised the long term future for shorter term profits. Eventually after a number of boardroom manoeuvres which it is easy to imagine Hedstrom finding tiresome he resigned on 24th March 1913. He received numerous offers of employment from other motorcycle manufacturers but he refused them all, preferring to live at his Connecticut home. Many people believe that in many ways the glory days for Indian were over despite the company's production of some superb motorcycles and considerable racing success after this date. Despite this the year saw the company's highest output – just short of 32,000 bikes – of which a quarter were exported.

Another name from Indian's recent past and another engineer, Charles B. Franklin, the TT rider (see Chapter Seven) was recruited to Indian's Engineering Department in 1914. The First World War brought mixed fortunes for the company. The upheaval in Europe affected much of their Export business but a number of Powerplus machines were supplied to the US Army. A dispute over the price of these military machines with the directors seems to have been the final straw for Hendee and he resigned from the company in 1916. The supply of military machines was not a particularly financially sound move for Indian but it did force them to appraise the cost effectiveness of their operations. They decided to sub-contract some of the component manufacture to specialists. This culminated in their selling the Hendeeville Plant to one of these specialists, The Moore Drop Forge Company.

While this helped it perhaps wasn't enough as sales of motorcycles declined in the early Twenties and the company made a loss. In 1922

Above
The single cylinder is inclined slightly backwards and takes the place of much of the seat post

Right
The carburettor is of Oscar Hedstrom's own design and bears his name – this type of carburettor was standard on Indians for several years

10

Indian suffered labour relations problems and the US domestic motorcycle market continued to shrink. Overseas this wasn't the case and there was still demand for American V-twins, perhaps nothing has changed in seventy years! The overall picture wasn't good and a loss of over $1 million didn't bode well for the future. Some changes of senior personnel left Frank Weschler in charge and under his leadership the company managed to turn the loss around and into a $200,000 profit for the next financial year. Around 50 per cent of production went overseas, especially to strong markets in Australia and South Africa. That November the company officially changed its name from the Hendee Manufacturing Company to The Indian Motocycle Company and in reward for his service Weschler was made President. Because of this a new optimism pervaded the company for 1924 although the motorcycle had definitely become a minority form of transport. In America at least cars were mass produced and cheap, they retailed at approximately the same price as a large capacity V-twin outfit. Fuel was cheap and of course cars offered better carrying capacity and weather protection. Despite this Indian racing successes were used to promote the make to the committed motorcycle enthusiasts of the day.

Above
The lubrication was total loss and a sight glass – seen here – allowed the rider to check on oil remaining in the tank

Above right
The camel-back tank, so known because of its distinctive shape was internally divided to allow it to contain both petrol and oil

Right
The early Indians clearly show their bicycle origins through the use of the diamond frame

Later Days

Circumstances outside Indian's control led to the loss of some export business. In 1925 in the UK, for example, a 33 per cent import tax was imposed on foreign bikes. This forced Billy Wells, the successful UK importer to cease trading within a couple of months. On the whole though there was continued prosperity for Indian although 1927 saw Weschler's resignation. It came about after a disagreement with the board of directors over the investment of company funds in new tooling and machinery. This demoralised a lot of the longstanding employees as it was generally accepted that Indian's survival in the recent difficult years was down to Weschler.

Louis J. Bauer, an Ohio industrialist, took over. Diversification was what he and the majority shareholders had in mind and their first experiment was with a small car, not unlike an Austin 7.

Growing inflation led to pressure on the US Government to introduce import tariffs. President Hoover succumbed to the pressure and tariffs on imports, including foodstuffs, were levied. The countries affected, such as Australia and New Zealand, retaliated by introducing tariffs of their own on imports including motor vehicles. Exports of Indian Motorcycles to these countries ended very shortly afterwards.

In the last years of the decade the company was still attempting to diversify and started manufacturing shock absorbers for cars. There were numerous complaints about the bikes dispatched to dealers at this time. The dealers were starting to feel neglected and saw the fact that bikes were supplied with parts missing as evidence of this neglect. Apparently it wasn't uncommon for bikes to arrive from the factory with clutch plates or gaskets missing and many dealers relinquished their franchises at this time. Bauer realised he'd not done the best of jobs for Indian and resigned in June 1929. A new President, J. Russell Waite, was elected. Cash was in short supply and soon bikes were only supplied to dealers on a 'Cash on Delivery' basis. Certain of the executives were looking at the fledgling aircraft industry with a view to becoming involved. Ultimately though, the company started producing outboard motors for boats. One of these engines was known as the Indian Arrow. E. Paul Du Pont, another wealthy industrialist, acquired control of Indian for reasons that had as much to do with his interest in aeroplanes as motorcycles. Du Pont called a halt to the outboard motor production.

After this unsuccessful period the company reverted solely to motorcycle production with a range that saw only minor upgrades. For '32 the Indian range was streamlined to what in effect were three similar machines with

Indian introduced the leaf spring suspension front fork in 1910 to replace the bicycle style forks. This feature would endure on Indians until 1946

An unrestored 1911 single cylinder
Indian, note that the 'loop' frame has
replaced the earlier bicycle style
diamond frame, which it did in 1909

different engines. Such a policy had a beneficial effect on the production
costs. The Chief, the Four and the Standard Scout engines were all fitted
to similar frames with common cycle parts. Despite this move the
company still made a loss for that financial year. It wasn't entirely
Indian's fault as the world was plunged into the Great Depression. The
end result was that Indian's workforce had to be reduced and further
savings made during the manufacture of motorcycles. The Depression also
led to there being a glut of cheap secondhand cars available so any
benefits that the motorcycle could offer in terms of economy was negated.
Sales of new machines were particularly hard hit. For 1933 Indian
produced less than 2,000 machines, somewhere near 20 per cent of the
entire US production that year.

Things picked up somewhat in 1934 and Indian produced more than
2,000 machines. The company also hired G.Briggs Weaver as Chief
Engineer to fill the vacancy left by Charles B. Franklin's death in 1932.
One of his first tasks was to redesign the range slightly. The company still
recorded a loss for the financial year but announced a comprehensive
range, sales of which, it hoped, would be boosted by racing success. The

likes of Rody Rodenburg and Ed Kretz brought about that success (See Chapter 7).

The impending war in Europe helped the economic climate to improve and worldwide military business was sought. Civilian production was largely suspended after the Japanese attack on Pearl Harbour and the American manufacturing industry as a whole went completely onto a war footing. Indian received an Army-Navy Production Award in 1944 for their achievement in production of military equipment at a special ceremony at the Springfield factory.

There were postwar changes at Indian with DuPont elected to a new post as the Chairman of Indian's board of directors and a number of new senior appointments being made. In October 1945 industrialist, Ralph Burton Rogers acquired a majority of Indian shares, taking formal control on the 1st November. Rogers wanted to continue motorcycle production but faced an uphill struggle. The Marshall Plan, intended to aid the reconstruction of Europe, somewhat handicapped domestic US manufacturers who were to be faced with imported competitors. Rogers, Briggs Weaver and others were convinced of the viability of lightweight motorcycles designed along more European lines and set about manufacturing such machines. Alongside such a project, the Chief was being produced again with a number of significant improvements which are detailed elsewhere in this book.

There were reliability problems with the vertical singles and twins that led to a great deal of dissension between the directors and the dealers. Dealers wanted the prewar Sport Scout to be put into production again as they considered it a better machine than the postwar verticals which were costing the factory more to produce than anticipated. One coalition of dealers actually went as far as to travel to the factory to protest to the directors in person. The financial state of the company can at best be described as precarious and Rogers travelled to England to look into the prospects of importing British bikes. He formed a deal with John Brockhouse of Brockhouse Ltd, an engineering company from Southport, England. Brockhouse also invested in The Indian Motocycle Company and sent one of his engineering staff to the USA to assist in improving Indian's own vertical twins and singles. The twins publicly lost face at the famous Laconia, New Hampshire races. Fifty vertical twins were entered by both factory and privateer riders. Sadly all dropped out with magneto failures while a dozen or so prewar Sport Scouts romped home. This wasn't to be the death knell for Indian's competitors to the imports but a currency devaluation across the Atlantic was. The British Government devalued the pound and it therefore lowered the retail price of the British imports by some 20 per cent. Indian weren't in a position to compete with such prices in the lightweight motorcycle market. This left Indian in a

New for 1911, the year that this machine was built, as an option was the 'free' engine. In other words, a clutch was available. This machine has one and the beginnings of what has become known as the primary cover can be seen

In 1911 Indian had eight different machines in their range including different types of ignition and capacities

desperate financial position and various financial meetings took place. Indian as a company was split into two portions; one was to continue the import of British motorcycles while the other was to continue the manufacture of Indian motorcycles. The former was known as the Indian Sales Corporation and the latter under control of the Titeflex Corporation who were owned by The Atlas Corporation. Many disgruntled dealers had resigned their Indian franchises and had taken on those for British motorcycles although a few dealers were still selling the heavyweight Chiefs. There were significant numbers of enthusiasts of these, and the competing Harley Davidson machines, who still remained loyal to the American products. Titeflex abandoned production of vertical twins in 1952 after they had made less than five hundred. The small volume of motorcycle production that Titeflex were now involved with didn't show satisfactory profits and in 1953 production of Chiefs was halted. The glory days of the 'Iron Redskins' were over.

Indian Singles

The first production motorcycle made by The Hendee Manufacturing Co was a single cylinder machine, with a displacement of 13 cu in (approximately 225cc) that produced 1.75 horsepower. Its cycle parts were essentially pedal cycle derived; a pedal cycle style diamond frame with wood rimmed wheels of 28in diameter. The engine was fitted into the frame with its crankcase just above the chain wheel bracket. The cylinder was inclined towards the rear of the frame and took the place of the majority of the seat tube. The remaining portion of this tube was bolted to the cylinder head. The ignition timing system ran off the crankshaft, as the French De Dion engines, power was supplied by three dry cell batteries carried in a tube mounted on the frame's front down tube. The coil was mounted below the front downtube. Fuel and oil were carried in an asymmetrical tank. This style of tank later became known as the 'camel back' because of the way in which it sat atop the rear mudguard. The bottom of the tank was curved to match the radius of the mudguard and the top was also curved. Inside the tank was divided into two compartments, one for petrol and a smaller one for oil. The lubrication was a gravity fed total loss system and a sight glass enabled the rider to check on the oil remaining aboard the machine – it was necessary to top up the oil every ten miles. A small canister type silencer was bolted to the frame under the bottom bracket. The engine controls were simply two levers mounted on the cross bar of the frame which moved a series of linked rods. One was for advancing and retarding the ignition while the other controlled the throttle valve on a carb of Hedstrom's own design.

Transmission was by means of a double chain arrangement – one ran from the engine sprocket to the countershaft in the hanger bracket which also carried the pedal chainwheel. Both chains were carried to sprockets on either side of the rear hub. The drive was permanently engaged as there was no clutch and a coaster brake was operated by back pedalling. The machine weighed 98lbs and was capable of speeds of up to 25 mph. Starting was accomplished by pedalling a short distance with the exhaust valve lifted, when sufficient speed to crank the engine was attained the valve was dropped. The motorcycles produced through to 1904 were of essentially the same design with only detail changes and improvements but for the 1905 sales year a number of improvements were to be incorporated. Incidentally the famous red paint that was to become known as 'Indian Red' was offered for the first time in 1904.

Improvements for 1905 included the fitting of a cartridge front fork in place of the solid bicycle type to give a more comfortable ride. The spring

Above right
George Twine's 1913 Standard. Of particular interest is the 'cradle spring frame' made available in this year as an option. Indian were the first manufacturer in the world to offer rear suspension on a motorcycle

Right
This bike was found as a runner in Maine twenty years ago and restored by George. He had to make the petrol tank which perhaps gives a clue as to the amount of work that has gone into the restoration

action could be adjusted by the slackening and repositioning of the bolt that secured the spring. Twist grip controls were fitted allowing the rider to better control the engine without taking his hands off the handlebars. This was seen as a significant safety improvement given the poor state of the roads of the day. While Hedstrom retained the same displacement and horse power rating the speed and power of the 1905 model was increased by improved camshaft operation and better valve timing. Steel cylinder barrels were lathe manufactured from solid forgings and contributed greatly to increased engine longevity.

The 1906 and 1907 models benefited from successive detail changes which saw a range of saddles being offered as options allowing a customer to choose the one that most suited his weight. The engine was made yet more powerful with a claimed output of 2.75 hp although weight increased to 115lbs. Acetylene lighting was now available as an option.

Like 1905 before it, 1908 saw substantial changes made to the products available from the Hendee Manufacturing Company. Singles of three displacements were offered; 19, 27 and 30 cu ins Magneto ignition was also available.

The motorcycle, as it is now referred to, was at this time still evolving into a form of mass transportation and as it had been derived from the bicycle it utilised the frame of such machines. The 'autocycle' as it was then known or the 'motocycle' (no r) as Indian had it used their bicycle frame. There was no collective thought on the appearance of early motorcycles – the Indian looked as it did because of its bicycle origins and the fact that it worked well in that configuration. Other manufacturers were designing motorcycle frames from a different starting point and were, therefore, not using bicycle style frames. They had devised what is generally referred to as the 'loop' frame, so called because the frame looped around the engine. This frame offered a number of advantages of which a lower centre of gravity and an aesthetically pleasing shape were not the least. The potential of this new design was not lost on Oscar Hedstrom and in 1909 Indian offered a loop frame design of motorcycle. They did retain diamond frames in their range for the 1909 sales year but as they were available at cheaper prices than the loop framed model it is generally accepted that this was to enable the company to use up its remaining stocks of diamond frames and compatible components.

Along with the move to a loop frame Indian offered a variety of options on their bikes – different handlebars, seats, footrests and even tyres were available – and a belt drive single was debuted in response to dealer pressure. Indian were pioneering chain drive while other manufacturers were still using belts and some dealers felt this was costing the company sales. Whether this was so or not is a matter for some debate and Indian found that sales of the belt drive model didn't justify the production costs

The carburettor of Oscar Hedstrom's own design was fitted to all Indians until 1915 when a cheaper Schebler item was used. The Schebler was already in use on other motorcycles of the day

The 30.50 cubic inch displacement engine drives the back wheel through a single speed transmission

and discontinued the model at the year's end.

1910 was another year where significant changes were introduced to Indian motorcycles; one was the introduction of the leaf spring front forks something that would endure into the mid-forties. A short front mudguard was utilised in conjunction with the leaf spring as it was considered that the spring was sufficient to deflect the road grime. Also introduced as new for 1910 were a 'free' engine and a two-speed gearbox. The former of these was simply the introduction of a clutch between engine and transmission while the second is self explanatory. Footboards became a standard fitting on several models. Indian also produced a sophisticated belt drive single but like its predecessor sales didn't justify its production and it too was dropped after only a single year.

In 1911 the free engine was sold in considerable numbers as an option on the 4hp single and the 7hp twin. Indian made another attempt to sell belt drive machines this time marketing a more conventional machine. With this the company had a line up of eight models in its range. This included motorcycles with a choice of battery or magneto ignition.

After their 1911 TT win (see the Racing chapter) Indian capitalised on their success with the introduction of Tourist Trophy models. These bikes, both singles and twins, featured kickstarters which were kicked forwards. They also featured double rear brakes; a foot operated external contracting band and a hand lever operated expanding shoe. This was partially occasioned by the legal requirement for motorcycles sold in Great Britain to have two independent brakes. Obviously Indian's British business was worth protecting. The 1912 model line up was reduced by two when the battery ignition option was withdrawn but the free engine option was now available across the range as was a two speed transmission. Also withdrawn were belt drive bikes once existing stocks were sold. Indian Red was the standard finish.

The V-twin engine was rapidly gaining in popularity as the more powerful characteristics of this configuration made it more suitable to America's long distances and then poor roads. As a result 90 per cent of the company's 1913 production was of V-twins. This didn't however sound the death knell for Indian singles but they would never enjoy significant popularity again.

Indian believed that there was a market for a small capacity, lightweight motorcycle and launched the model K Featherweight in 1916. This 221cc two stroke machine produced 2.5 bhp. It had an external flywheel, magneto ignition and was lubricated by an oil spray at a rate of one pint per 25 miles. The gearbox was three speed and both primary and final drive were by means of chains. The bike was capable of 35 mph and 100 mpg. Sadly it wasn't a success and was dropped after only one year in production.

Innovative thinking from Indian culminated in the launch of the 'cradle spring' frame in 1913, the year in which the company accounted for more than 42 per cent of total domestic production, a time when Indian could indisputably claim to be the largest motorcycle manufacturer in the world.. The Hendee manufacturing Company were the first motorcycle manufacturer anywhere to offer what has since become known as swinging arm rear suspension. The rear portion of the lower frame tubes were designed to pivot behind the gearbox and springing was provided by quarter elliptic springs in a similar way to the front suspension, although at the rear there were a pair of springs. This system worked in place of the latterly more common telescopic dampers. The rigid frame remained available as an option.

A later non standard carburettor has been fitted to this machine in place of the original Hedstrom type

Indian Twins

It was not until 1907 that Indian had V-twins in their model line up and in that year they offered two; a 38.61 cu in roadster and a 60 cu in racer. Like the singles the twins had an inlet over exhaust valve f-head configuration with mechanically opened exhaust valves. The inlet valve was either mechanically actuated although an optional suction valve system was available. On the roadster bore and stroke were 2.75 and 3.25 in respectively. Lubrication was total loss and the transmission was single speed with a chain final drive. The twin was capable of 60 mph although the rear sprocket affected this and there were different ones available for racing or road use. Otherwise the early Indian twins were similar to the singles, many of the cycle parts were common to both models. It was from 1913 onwards though that Indian's V-twins would come to the fore.

An unusual twin was the result of Indian's attempts to market a motorcycle for the lightweight market. The Model O was introduced in 1917 as a replacement for the lightweight Model K single. Although the O Model was a twin it was unusual in that it was a horizontally opposed twin, it was also of small displacement. Unlike the popular BMW Boxer twin the Model O's cylinders ran fore and aft along the frame in the manner of the Douglas then manufactured, and popular, in England. Induction and exhaust, on the engine which had been designed by Charles B. Franklin, were by means of side valves, Bore and stroke were 2 in and 2.5 in respectively which gave a total displacement of 15.7 cu in (257.3 cc). Ignition was by means of a magneto. Much of the engine was made of cast iron including the pistons and cylinders. The double throw crankshaft was a single forging that turned in bronze bearings, at one end was attached an external flywheel. Transmission was a three speed handshifted with a dry clutch. Both primary and final drive were by means of chains. This lightweight was claimed to be capable of 45 mph and 80 mpg.

The Model O was acknowledged as being easy to start, pleasant to ride and for providing acceptable performance. It wasn't however, a commercial success for a number of reasons. Many dealers preferred the proven V-twin engined machines and promoted these at the expense of the lightweights. Some dealers also considered them too expensive, a catalogue price of $180 was dearer than a used car of the time. Production was discontinued in 1919.

In 1916 Indian introduced the Powerplus engine and it was radically different from Hedstrom's f-head machines as it was a sidevalve design. A displacement of 60.88 cu in was achieved by a 3.125 in bore and a 3.99 in stroke. The design deliberately had a long stroke because they wanted to

Above right
Although the cradle spring frame was available in 1913 as an additional cost option this bike didn't benefit from the innovation and came in a rigid frame

Right
Although this motorcycle is a 1913 model it has been fitted with a later 3-speed transmission. This was a common modification to uprate earlier clutchless machines

build an engine that produced a lot of torque. The cylinders were arranged in a 42 degree V. The cylinders and cylinder heads were one piece castings, access to the valves was by means of removable caps and access to allow priming of the cylinders with a syringe was to aid cold weather starting. There were two valves per cylinder all operated from a
single camshaft. The Powerplus was nominally designed by Charles Gustafson senior although much of the work was done by Charles B. Franklin. Certain of the engine's functions such as a total loss oil system were carried over from the F-head and some of the components there was much new about the engine. The Hedstrom carb was replaced by a Schebler item which was cheaper. Franklin went on to design the Scout and Chief models and to some extent turned his attention away from the Powerplus. This meant that the changes made to it for the rest of its production run were minimal.

Despite a total loss lubrication system many aspects of the Powerplus were modern by comparison. The transmission was three speed with a foot operated dry clutch and a hand shifted gear lever. Primary drive was by a chain as was the final drive. Ignition was magneto and the 410 lb motorcycle was capable of an estimated 60 mph. It returned 35 mpg and used oil at a rate of a pint every 50 miles. Production of this model finally ceased in 1924.

The Indian Scout

One of Franklin's best designs was the 37 cu in Indian Scout, it was a sidevalve V-twin that first appeared in late 1919 ready for the 1920 sales season. It quickly established itself as a solid reliable motorcycle and engine longevity became a major selling point; 'You can't wear out an Indian Scout' went the slogan. The primary drive on the Scout was by means of helical gears running in oil instead of the more common chain. Once again changes were minimal to this successful bike on which a number of records were set, the helical gear primary remained a standard feature until 1933 for example. This type of primary drive was noisy but enormously long lasting. Removable cylinder heads appeared in 1925 and a 45 cu in variant appeared in 1927. The extra displacement was obtained by increasing the bore from 2.75 in to 2.875 in and increasing the stroke from 3.0625 in to 3.5 in. This was followed by the Model 101 Scout in 1928.

A problem Indian faced with regard to export sales was that of foreign competition. In the early twenties however a couple of manufacturers took the idea a step further; in Osaka, a Japanese company produced exact copies of the 1921 Scout. They are reported to have been indistinguishable from the American originals. A similar thing happened in Germany where

90 per cent of Indian's 1913 production was of V-twin engined motorcycles of which this was one

a motorcyclist, named Max Bernhardt, produced the Mabeco. This was another replica of the Scout that first appeared in 1923. The Mabeco was produced in Berlin and endured for several years. In 1927 Reinhold Kleiber took over production and went on to build copies of the 1927 45 cu in Scout. The Mabeco had a German made carburettor, a Siemens magneto and the screw threads were metric.

The 101 Scout was something of a new motorcycle with an increased wheelbase; extended approximately 3 in to make it 57.125 in, in total. A lower seat height and a more slender petrol tank added to the more graceful overall appearance. The 45 cu in sidevalve engine produced 18 bhp and was capable of over 70 mph. Many riders and dealers thought that the 101 Scout was Indian's best ever motorcycle and there was an outcry when the 101 Scout was discontinued in 1931. The replacement Scout was a 45 cu in engine in a Chief frame. A smaller displacement motorcycle with a 30 cu in engine was launched partway through the year and referred to as a Scout Pony. This was subsequently followed by the Junior Scout and Thirty-Fifty models, the latter name referring to the cylinder displacement of 30.5 cu in (500cc).

In 1933 Indian announced dry sump lubrication on all its twins; a major step forward and ahead of major rivals Harley Davidson. A 45 cu in version of the Scout Pony, known as the Motoplane, was launched but the lighter frame of the Scout Pony wasn't up to the extra power unlike the Scout engine in a Chief frame. Racers and enthusiasts wanted a true sports

Above
This V-twin was registered new in Derbyshire, England in 1913

Right
The frequent maintenance required by early motorcycles is indicated by the necessary spares carried in this tank top box

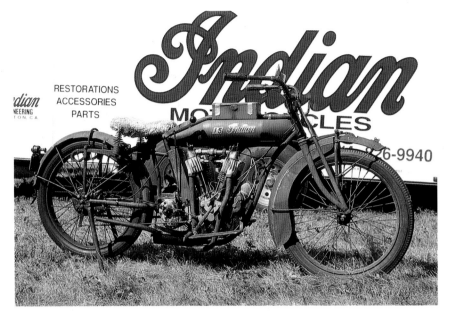

Left
Indian capitalised on their Isle of Man TT victory and marketed motorcycles known as Tourist Trophy models. This unrestored 1915 V-Twin was one, it now belongs to Max Bubeck

Right
The 1915 F-head engine configuration such as this one was carried over into the Powerplus a year later

bike to replace the 101 Scout and responding to this Indian launched the Sport Scout in 1934. English style girder forks added to the sporting appeal of this model. The Sport Scout used what the factory termed a `keystone' frame, it bolted together and the engine took the place of much of the bottom rails. The Sport Scout was also blessed with redesigned graceful mudguards. To differentiate it from the other Scout the existing model became known as the Standard Scout. Also for '34 Indian reverted back to a chain drive primary that ran in an oil bath.

1935 saw the Standard Scout fitted with redesigned mudguards, similar to those on the Sport Scout but with deeper valances. Minor changes were made to the ignition system for 1936 and the Standard Scout was renamed the Scout 45. Most of the changes made to the Sport Scout from then on mirrored those made to the Chief including the changes made to the cylinder heads in 1938 to ensure cooler running. The big swoopy mudguards associated with the big postwar Chiefs were fitted to the 1940 Thirty-Fifty and Sport Scout models.

As Americans put on olive drab for the duration of World War II so too did the Sport Scout. Indian produced a 30.5 cu in version of the Sport Scout and designated it the Model 741. The US Government evaluated these motorcycles alongside the larger capacity Harley Davidson 45 and finally chose the latter for large contracts. Indian did sell numbers of the 741 to the allies and as a result they are plentiful in Europe, Australia and New Zealand. Postwar civilianised war surplus were a common sight on the roads in these countries. A number of 45 cu in versions, designated the Model 640-B, were sold to Canada and all were sidecar equipped.

The Indian 841

It is truly said that 'necessity is the mother of invention', it's also true that war speeds innovation. Indian obtained some military business in the early years of World War Two but sought further contracts. The US Government wanted shaft drive motorcycles so Indian set about developing one in an attempt to gain official contracts. They designed a machine that was innovative in a number of ways and yet still managed to use some existing components. Still a V-Twin, the 45 cu in engine was arranged across the frame in a manner that both Moto Guzzi and Honda would later emulate. The engine, despite its unusual orientation, used a number of standard Sport Scout engine parts. The machine was shaft driven and suspension was plunger at the rear. Up front suspension was effected by a girder fork set up. The handlebars were rubber mounted to reduce the vibration transmitted to the rider. Only 1000 of these bikes were built and it is thought that they saw no military service beyond evaluation. The reason for not putting the 841 into production was simply that the US Army preferred another innovation; the Jeep. The project wasn't a total loss though, a version of the forks and handlebars appeared on postwar Chiefs, as did the hubs and brakes.

Above
After the end of hostilities the military disposed of many of its motorcycles, as a result civilianised versions of the 741-B like this one were common in post years. This one has a pre-war Indian colour scheme

Right
A 500 cc version of the Sport Scout was supplied by Indian to a number of the allies involved in World War Two including Britain, Australia and New Zealand. As a result they're relatively plentiful outside the USA, this one is on display in Max Middelbosch's museum

Paul Pearce of Oxford, Michigan is the owner of this immaculately restored 1917 Model O Light Twin. US entry into the Great War in that year would of course have a profound effect upon the company. It offered more motorcycles for the war effort than all the other domestic manufacturers put together; and inflation would eat into the already suffocatingly tight margins on military production

Above

The O Model is unusual for an Indian Twin in that it is of horizontally opposed configuration and of only 257cc displacement

Left

The sidevalve twin featured an external flywheel and was noted for its smoothness

It is possible to date the machine is 1917 Model O because it has the coil sprung cartridge front suspension. The 1918 and 1919 models were fitted the more common leaf spring set up

Above

The Powerplus was a particularly noteworthy motorcycle when it was introduced in 1916 because it was Indian's first sidevalve V-twin

Left

This 1918 model features the 1000 cc sidevalve engine that, together with the three speed transmission, was to be the foundation of Indian's later success

Right

The pushrod tubes for the sidevalves can be clearly seen in this photo as can the fact that the cylinder heads and barrels were a single casting

The Indian Chief

The Indian Chief is probably the motorcycle everyone thinks of when Indians are mentioned because of its distinctive lines, big valanced mudguards, enclosed chain guard and opulent fittings. That particular Indian Chief, though, is almost the last of the line and was produced in the postwar years. The Chief as a model designation goes right back to 1922. In that year Indian put on sale a 61 cu in V-twin motorcycle that was another of Charles B. Franklin's designs. It was initially billed as a Big Scout and emphasis was put on its suitability for sidecar work. Perhaps surprisingly Indian had seen fit to drop the cradle spring frame (see Chapter 3). Several of the parts and ideas were carried over from the Powerplus. This included a number of transmission parts and the one piece cylinder and cylinder head castings. Access to the valves was by means of removable caps. The Chief stayed with this basic configuration through 1923 when a larger displacement Chief was introduced. This bike was dubbed the Big Chief and displaced 74 cu inches. The process of updating details on the bikes continued through the turn of the decade. One important improvement was the fitment of removable cylinder heads on '25 model Chiefs. Other changes such as the use of cast aluminium petrol tanks on 1930 Chiefs weren't particularly successful and were dropped after only one year. Apparently the factory experienced problems with porosity of these tanks and had to reject too many tanks to make it viable.

In 1932 came a completely new look Chief, with a taller profile. The new models had longer front forks and a taller frame to accommodate them. The fuel tank was redesigned too, no longer did it sit between the frame tubes and was much more elegant. The ignition system was modified as well, a wasted spark battery ignition set up being employed. The lubrication system remained total loss. Again the model endured several sales years with only detail improvements and changes; a variety of paint schemes reflecting Du Pont's ownership of Indian, new tail lights and restyled mudguards.

An option for the 1935 Chief was the Y-motor which had larger cooling fins on the barrels and aluminium cylinder heads. A four speed transmission was also an available option. For 1936 the ignition system was changed again, this time to a coil system. An Autolite distributor became standard and the magneto ignition system was only available if specifically ordered. In 1937 the gear lever was moved from the rear of the tank to the front necessitating a more complex linkage. Another 'improvement' that wasn't and therefore only lasted one year was the provision of an external oil line from the crankcase to the pushrods. It was dropped because of the excessive oil that collected on the exterior of the engine. This didn't stop the experimentation though, and a new oil pump

Above
The Big Chief, a 1200cc machine, made its appearance in 1923. This example belongs to Donald Haras, seen here with his motorcycle, and dates from the second year of production

Above right
The Big Chief still had much in common with the Powerplus including the one piece cylinder castings

Right
The Mabeco was a copy of an Indian Motorcycle manufactured in Berlin, Germany. Note that the 'Mabeco' name on the alloy casting is in Indian script

that incorporated the distributor was instituted to improve matters.

The World's Fair took place in 1939 and for this year most of the improvements were cosmetic. New paint schemes were unveiled on bikes that carried a lot more chrome plate than previously. A new rear bumper, air cleaner and upswept exhaust all benefited from the chroming process. The hugely valanced – or skirted – mudguards referred to at the beginning of this chapter first appeared in 1940 right across Indian's range. A more major change also made in this year was the provision of plunger rear suspension in the previously rigid machines. The plunger is essentially a set of coil springs above and below the axle on each side of the wheel. The upper ones absorb the bump while the lower ones 'damp' the recoil. For this year Indian experimented with a spin-off cartridge style oil filter but did experience cold weather oil flow problems.

In exactly the same way as Indian relied on Police contracts for a proportion of their sales, military business was seen as similarly desirable. It wasn't necessarily especially profitable but increased the volume of machines produced and so ensured some economies of scale. Indian sought to supply motorcycles to both US Forces and the allies. In October 1939 an order for 5,000 military Chiefs with sidecars for the French Army was seen as a welcome windfall. The order was completed in March 1940 and approximately half the machines dispatched to New York. Here they were loaded aboard the SS Hanseatic Star for delivery to France. Unfortunately the ship never arrived being lost at sea en route, a victim of the U-Boats in the Battle of the Atlantic. Military Chiefs were simply basic versions of

The crankcase and primary cover were originally painted Indian Red as on this Big Chief

the 1940-specification civilian machines. They were lacking in chrome and valanced mudguards and uniformly finished in olive drab.

Postwar the Indian line up consisted only of the 74 cu in Chief. One of the most obvious updates to the machine was the use of girder forks. This softened the ride and more than doubled axle travel. The rear plunger set up was also softened to keep up with the extra front suspension. The 1947 models were simply refined '46 models with different tank emblems, a redesigned kickstart cover and chainguard. A new style of clutch release bearing was an 'improvement' that was dropped for 1948. For that year there were alterations to the exhausts.

In terms of Chief production 1949 was a quiet year for Indian because the company was in the process of moving to a single storey factory, in fact back to the founders original building, mainly for reasons connected with the production of singles and twins (Chapter 6).

The Chief was back in 1950 with an enlarged engine displacement of 80 cu in and telescopic forks. Less obviously it also had a right hand throttle twistgrip, the first production Indian to do so. The increased displacement

Above
The large displacement engine made the Big Chief suitable for use with a sidecar. Brett Colson's 1928 model is seen here with a Goulding sidecar

Above right
New for 1928 were the lower overall seat height and a more slender tank that gave the bike its graceful lines

Right
The sidevalve 42 degree V-twin engine displaced 45 cubic inches and produced an estimated 18 bhp. Transmission was three speed

Above

The showroom stock 101 Scout was capable of over 70 mph and more when tuned. This coupled with good handling characteristics meant it was a popular bike that won the affection of many Indian riders

Right

A 1928 101 Scout restored by Tony Leenes, the longer wheelbase ensured the excellent handling for which the 101 Scout has become legend

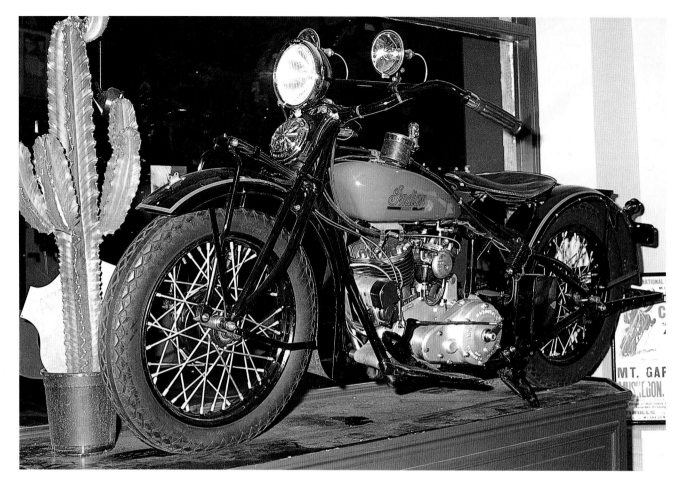

had been obtained by lengthening the stroke from just under 4.5 in to almost 5". 1951 Chiefs were the same as the year previous but for 1952 there were a number of alterations. The front mudguard was redesigned for the second time making it slightly smaller than the previous ones. The saddle was mounted rigidly and the exhaust was rerouted. Redesigned petrol tanks and chainguard were standard and an engine cowling was available for the first time. A fork cowl was fitted in 1953. In these latter years a number of changes were made for both practical and cost reasons. Cheaper English Amal carburettors were used in place of Linkerts and Amal handlebar controls were used. Indian used their surplus vertical twin silencers on the last Chiefs. The total output of Chiefs for 1952 and '53 is estimated to have been only 1300 machines.

Above
A restored 1931 Scout on display in the window of the Edinburgh Indian dealer, Motolux, run by Alan Forbes

Right
The Sport Scout, introduced in 1934, used a bolt together frame termed a `keystone' frame by the factory. Many Sport Scouts were used for racing

Above
Elmer Lower starting his perfectly restored 1936 Chief at The Daytona International Speedway. This bike has no extra chrome and only accessories that were offered by Indian themselves

Top
This 1937 Chief has been retro-fitted with later cylinder heads and barrels

Left
The dashboard of the 1937 Indian Chief

Above
For 1941 the Sport Scout was fitted with the hugely valanced mudguards more typical of the postwar Chiefs. Also new for '41 was the plunger rear suspension

Right
The 1940 and 41 Sport Scouts featured redesigned cylinder head and barrels to assist in cooler running and the headstock angle was altered to improve handling

Above

The rear suspension system is acknowledged to have worked, there are those who feel it could have been better and those who feel that there wasn't much difference in handling between the bikes with rear suspension and the earlier rigid models

Left

Another of Elmer Lowe's restorations; a Fallon Brown and Indian Red 1941 Sport Scout. It is extremely unusual to find a Sport Scout out-of-the-crate today, because the racers would invariably strip the machines down. There was only one more year of production after this example first rolled out

Left

Neil Grieve from Edinburgh astride the 1947 Chief he restored, rebuilding every moving part along the way

Below

1947 was the first year in which the Indian's illuminated head sidelight was fitted on Chiefs. That year also welcomed back chrome, with the improvement in supply after war shortages

The hugely valanced mudguards were redesigned three times in the last years of postwar Chiefs but new specifically for 1947 the name Indian appeared in script on the tank

Above

The first postwar Chiefs are easily identified by their girder fork assembly as seen on this fine motorcycle which also sports typical period saddlebags and windscreen

Left

The 74 cu in sidevalve engine was the only displacement Indian made available in the immediate postwar years

Above
Ken Young has owned this 1947 Chief for the past twenty years and describes it as a 'good, honest motorcycle'

Right
The 74 cu in engine as it was known actually displaced 73.62 cu in, produced an estimated 40 bhp and was good for 85 mph in stock form. It would return an average of 40 mpg

Overleaf
This black Chief is equipped with a buddy seat but can be dated as a '47 model because of its script tank emblem, Indian's head sidelight and girder forks

Above

An owner checking the engine idling of his Chief in the Daytona sunshine. The streamlined rear mudguard is, of course, obscured by the saddlebags

Below

During the production of postwar Chiefs detail changes were made to components like the oil pump and distributor drive

Overleaf

The primary drive on 1946 to 1948 Chiefs was an endless chain running in a cast aluminium oil filled primary case

Above

1948 was the last year of girder forks although for 1949 few Chiefs were produced due to the company moving factories

Left

A late-forties Chief being ridden on Daytona Beach. The 1948 Chief gained a big new speedo in a stamped-steel dash and an aluminium oil pump

Left

For 1950 the 74 cu in Chief was enlarged to 80 cu in (1300cc) by means of increasing the stroke by half an inch. The tank emblem was redesigned to proclaim this increase in size

Above

The second thing that was new for 1950 was the fitment of telescopic forks seen on this Chief. Harley-Davidson's Hydra-Glide forks were introduced the year before

Right

The telescopic forks didn't detract from the styling of the Indian Chief although the front mudguard was subtly redesigned becoming slightly smaller than previously

Above

The heart shaped dash panel had been dropped in 1948 in favour of the more oval shaped item seen here on this late Chief

Above

One of the last significant changes made to the styling of the Chiefs was the fitment of an engine cowling on 1952 and '53 models

Left

A fork cowling was also fitted to 1952 and '53 Chiefs. The rear suspension was softened to make it more compatible with the suspension of the telescopic forks up front

Right

American Motorcycling tested the 80 cu in Indian Chief in December 1950 and reported that the engine produced 50 bhp at 4800 rpm

Above
The red generator warning light seen on this dashboard had replaced the ammeter in 1948

Right
The tank detail of a Blackhawk Chief – the models were tagged `Blackhawk' with the introduction of the telescopic forks. A name wasn't going to keep the company afloat: the 1952-3 Chiefs were restyled with new fuel tanks, seats and fenders, but a cruel observer may have likened the exercise to a trout thrashing on a line

Indian Fours

Indian became involved in the manufacture of in-line four cylinder motorcycles at the beginnings of the Depression when in 1927 they purchased the rights and tooling for the Ace Motorcycle. The Ace had come about from the collaboration of two Scottish born brothers, Will and Tom Henderson. They'd manufactured quality motorcycles under their own surname from 1912 until 1917. They then sold their business to bicycle manufacturer and owner of Excelsior Motorcycles, Ignatz Schwinn. After a couple of years working for Schwinn the brothers resigned and set up on their own again although Tom's involvement was considerably less than before. For the two years after 1920 the company managed some production but in 1922 Will Henderson was killed in an accident while testing a motorcycle. Because of this and a precarious financial situation the company changed hands more than once, latterly to Indian.

Initially Indian made few changes to the Ace beyond using smaller wheels, painting the bikes Indian Red and renaming the model the Indian Ace. For 1929 a few more changes were made including the installation of the leaf sprung front fork assembly (already used on many Indian Motorcycles) and fitting a redesigned petrol tank. The 1929 model Ace was named the Indian Four Model 401. Later a twin downtube frame

Left
Alan Forbes at speed on his 1931 Indian Four. This motorcycle formerly belonged to the original Edinburgh dealer and appeared in Indian literature from the thirties

Right
1931 was the last year that the 'between the rails' fuel tank was fitted. The 1265cc engine produced sufficient torque for the rider to be able to move off in top gear

When the bike was discovered by Alan Forbes it had been garaged since 1939. Alan is seen here wearing the correct period costume to match the bike

replaced the original Ace's single downtube design as a way of reducing
vibration.

The Model 402 was the replacement for the 401 and introduced in May
1929 with redesigned engine internals. The Ace had a three bearing
crankshaft and this had been upgraded to a five bearing one. The valve
arrangement the proven inlet over exhaust configuration, bore and stroke
were 2.75 and 3.25" respectively which gave a total displacement of 77.21
cu in (1265cc). The engine was air cooled, returned an estimated 50 mpg
and was capable of over 70 mph. Production of this engine continued until
1936. Transmission was a three speed, shifted by means of a hand shifter
and a foot clutch. Between 1928 and 1942 the wheelbase grew from 59.5
in to 62 in, in fact the main changes to the Fours were to the cycle parts;
a new taller frame and forks in 1932. A total of 24 paint schemes were
listed for '34 and a restyle for 1935. It is worth pointing out that sales of
these machines were very limited – the total for 1934, for example, being
approximately 200 machines.

In 1935 Indian announced, for 1936, the Model 436 but it became
known as the `upside down' engine, an unfortunate tag which did little

for sales. What Indian's engineers had done was to redesign the engine and in doing so reversed the position of the valves so creating an exhaust over inlet design. It was said to increase the power output. The visual appearance of the `upside down' engine was far more cluttered than its more conventional predecessors due in the main to the position of the exhaust manifold on top of the engine. The upside down configuration was continued for 1937 as the Model 437 with the addition of twin carbs, it was known as the Sport Four.

For 1938 a revised engine configuration was introduced for the Four and known as the 438. Indian had returned to the more accepted inlet over exhaust F-head arrangement and also incorporated completely aluminium enclosed valve gear and automatic valve lubrication. The cylinders and aluminium heads were cast in pairs. The Four was vastly improved and in this guise continued into 1939 with only minor changes. For 1940 changes

Above
For 1938 the four was returned to the proven inlet over exhaust engine configuration

Right
The more accepted inlet-over-exhaust valve configuration is seen on this 1938 Model 438

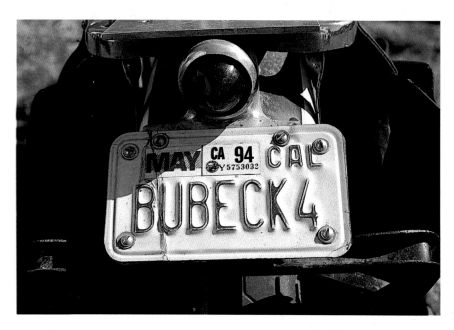

were made to the frame when plunger rear suspension was made available and 16 in wheels were now an option in place of the 18 in ones fitted so far. The wheels were partially concealed behind the large valanced mudguards. 1941 saw the addition of a chrome strip along the sides of the petrol tank. Production of Fours ended in March 1942 once outstanding law enforcement agency contracts were fulfilled, civilian production had been curtailed in December 1941 after the Japanese attack on Pearl Harbour.

The Four wasn't reintroduced after the war as sales had never been high. There were a variety of reasons; Fours were expensive motorcycles, dearer than cars of the time, partially as a result of the expense of manufacturing a complex engine. The early Fours had been marketed in a severe depression and many Fours, particularly the early Indian Aces and the 400 models, were prone to overheating and transmission problems. There were however two other projects based around four cylinder engines but neither really got beyond the prototype stage. The first was a shaft drive ohv in-line four of which two engines were built and tested.

Above
California vehicle licensing peculiarities allow such esoteric personal registrations. What could be more appropriate on Max Bubeck's Four?

Right
Max Bubeck, seen here riding his 1939 Four, has owned it since new

Wartime priorities, however, meant that the project was shelved. The Torque Four was a modular engine developed by the Torque Engineering Company. The concept was of an engine comprised of a number of common castings that could be bolted together to make a twin from two singles or a Four from four singles. It wasn't altogether successful although three Torque Fours are believed to have been assembled.

It would be the Japanese manufacturers in later years that would popularise four cylinder motorcycles, albeit in an across the frame configuration.

Indian Fours for 1940 and 41 were fitted with the valanced mudguards and plunger rear suspension that were fitted across Indian's range

Right

Production of Fours for civilian use was curtailed after the Japanese attack on Pearl Harbour. Once outstanding law enforcement agency contracts for Fours were completed in early 1942 production stopped altogether. This machine is one of the last Fours made

Indian Verticals

Ralph B. Rogers intended to have the vertical twin and single engined motorcycles built by Indian and marketed through the large existing dealer network. He instigated production and they went on sale in 1948 although they were billed as 1949 models. There were two bikes in the range; an overhead valve 220cc single and a similar 440cc twin. Designations were 149 and 249 respectively, both bikes had definite European styling including four speed gearboxes with foot shifters. Delays and production difficulties with these machines explains why few 1949 Chiefs were produced. The models didn't sell particularly well and for 1950 left over single cylinder models were renamed the Model 1150 and left over twins were offered as the Model 2250.

There were new models for 1950 though, the 500cc Model 250 Warrior and the Warrior TT. These too were overhead valve twin cylinder machines with a four speed transmission. The bikes sold only in small numbers but did endure in production until 1952. In that year the Titeflex Corporation abandoned their production.

An unusual accessory for the Verticals was the ski attachment intended to convert the motorcycle into a snowmobile. Dennis Bolduc is seen here demonstrating how it would attach

Above

The 1950 TT Warrior was the replacement for the Model 249 and its European styling clearly shows who Indian intended it to compete against

Left

This particular TT Warrior was restored to original by New Yorker Chris Lord who is seen here with his motorcycle

Indian were still interested in Police business and produced the Police Warrior. Apart from the flashing lights and tank decal one of the main changes from the Warrior TT was the use of footboards

A 'big frame' vertical twin. Indian produced this prototype in response to criticisms that the Warrior was physically too little. The frame was higher and longer than a standard TT Warrior. It had one off forks, a wider tank and wider mudguards but never went beyond the display stage. this one is now owned by Rick Brown

The vertical twins didn't achieve anything when raced at Laconia in 1948 but that hasn't stopped this one from being prepared for historic racing

Racing and Competition

As has been seen, competition success was important to Indian from their earliest days and one of their most successful wins was at the beginning of their second decade in business. Somewhat ironically for an American manufacturer it happened in Europe, at the 1911 Isle of Man TT.

The 1911 Isle of Man TT

Like the motorcycle itself the TT races were also in their infancy; in that year it was decided to use the 37.75 mile mountain circuit for the first time. Up until then the TT races were held on the much flatter 15.8 mile St John's circuit. Five laps of this new circuit was a demanding 187 mile task, meaning as it did that each motorcycle had amongst other things to climb the mountain road over Snaefell five times and pedalling gear had been banned. Also for the first time the racing was divided into Junior and Senior events. Then, as now, the race that would earn both manufacturer and rider most kudos was the Senior TT. The regulations for those early races appear to have incorporated something of an anomaly concerning cubic capacity. Single cylinder machines were allowed a maximum capacity of 500cc while twins were allowed a maximum of 585cc. The reason given for this was that twins were still considered unreliable and less mechanically efficient. While this wasn't strictly true that was how the rules stood when Indian fielded a team in the 1911 Senior TT. Hendee and Hedstrom decided to enter the 1911 event and in January of that year informed the UK importer Billy Wells, of their decision. Wells was enthusiastic and agreed to find riders to race alongside Hedstrom's nominee, Jake De Rosier.

The partners built twin cylinder 3.75 hp engines especially for the event to meet the required maximum displacement of 585cc. Each cylinder had a bore and stroke of 70 x 76 mm. Special small diameter flywheels and crankcase were manufactured. The remainder of the engine retained Indian's normal design — overhead inlet valves, side exhaust valves, mechanical oiling and removable cylinder heads. A two speed countershaft gearbox that had previously only been available on the 7 hp model and chain drive was utilised. The red painted bikes into which these engines were installed were also quite special, the gear control was moved forward close to the handlebars so that it could be more easily operated from the riding position adopted by racers. An extra oil tank was mounted behind the seat tube with a drip feed to the gearbox. Footpegs took the place of footboards and the clutch was converted to hand operation. A lever

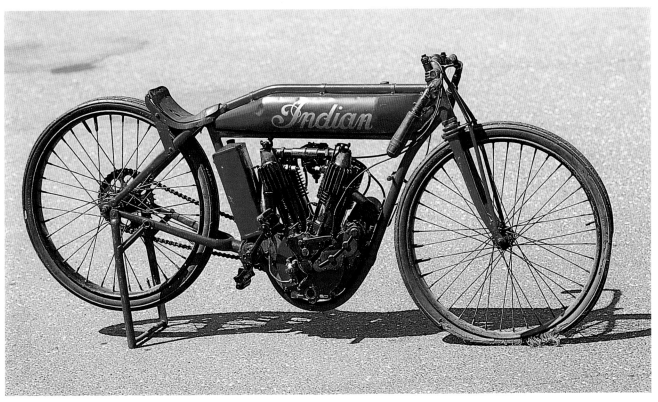

mounted alongside the petrol tank could be set in any position to allow the rider to slip the clutch as necessary. Brakes were foot controlled and twistgrips controlled the engine; the left one operated the throttle while the right advanced and retarded the ignition and lifted the valves. The front forks were leaf sprung and a silencer wasn't fitted as it wasn't required by the race regulations.

There were 64 machines entered in the Senior of which 24 were twins there were five Indians on the grid all backed by Billy Wells. Oliver Godfrey was one of Indian's star riders – a few months earlier he'd broken the single cylinder one hour record at Brooklands. Aforementioned American Jake De Rosier was another of their stars and another record breaker. The other three riders were Dubliner Charlie Franklin, Jimmy Alexander and Arthur Moorhouse, all experienced road racers. While the Indian team was a strong one there were formidable challenges from similarly experienced riders on Matchless and Scott machines – Charlie Collier and Frank Philipp respectively.

The racers were flagged away with the first lap taking the early leader de Rosier approximately 45 minutes to complete. He was half a minute

Left, above and overleaf
Four shots of the historic racing action at Daytona Speedway; Butch Baer on a 1926 Scout, Jim Smith on a 101 Scout, Butch Baer again, this time aboard a 1937 Sport Scout and finally a rare 1948 Daytona Sport Scout. This last bike is one of only fifty manufactured and the last of the Sport Scout line

ahead of Godfrey and Collier, Moorhouse was holding fourth and Franklin sixth. Things changed on the second lap when Collier took the Matchless into the lead leaving Godfrey in third. Alexander crashed, injuring his knee and damaging the twistgrip control of his Indian. There were more changes to come in lap three when Godfrey passed De Rosier and by the end of the lap the four Indians trailed the Matchless. Frank Philipp's Scott recorded the fastest lap of the day but he was well down the field. Things changed again in the fourth lap, Collier punctured a tyre and dropped down to third, De Rosier had mechanical problems and spent twenty minutes in Ramsey adjusting an exhaust valve and changing a spark plug. Franklin moved up to second although Godfrey was two minutes ahead of him. The excitement wasn't over and the final lap saw Collier again challenging for the lead passing Franklin and halving Godfrey's two minute lead. Despite this valiant ride Godfrey took the flag in Douglas. Collier was subsequently disqualified for refuelling in an unauthorised place meaning that second place fell to 25 year old Franklin and third to Moorhouse meaning a one, two, three win for Indian. De Rosier struggled home in 11th place but was also disqualified for using spares not carried on the machine. This triumph was the only time in the TT's long history where an American made motorcycle has won a race and was the first make of machine to ever take the top three places.

International Six Days Trial

Another competitive motorcycling event that was in its infancy in the first decade of the Twentieth Century but that was to continue and become significant in motorcycle sport was the International Six Days Trial (ISDT). It was first run in England in 1907 and known as The Thousand Mile Reliability Trial. It was held on public roads with checkpoints to be passed at fixed times, reliability rather than outright speed being the key to gaining a good position. T.K.Hastings – known as Teddy – from New York entered and won on an Indian Twin. He was no doubt aided by the reliability of the Hedstrom engine and the fact that the chain drive was an advantage especially in hilly areas. Teddy Hastings had borne the costs of his entry himself for the 1907 event but for 1908 Indian gave him some backing and he repeated the victory. Subsequently he became the Indian dealer in Melbourne, Australia.

Volney Davis

Volney Davis was an Indian enthusiast who made a noteworthy long distance ride in the summer of 1910. He rode a 5 hp Light Twin from San Francisco to New York City and then back to San Francisco. The

journey totalled 10,400 miles much of which was across trackless land, some was on dirt tracks and even some along railway lines. He reported few problems other than a number of punctures.

Willy Wright

Willy Wright, a ten year old, rode a 1911 single from the East coast to the West coast of America. He holds the record as the youngest rider ever to cross the continent. It is of course unlikely that this record will ever be broken because of the various US traffic laws that specify minimum ages for the riders of motorcycles.

Cannonball Baker

Erwin G. Baker was born in Indianapolis in 1882. He was something of an athlete and consciously only ate nutritious food and abstained from both alcohol and tobacco. His physical prowess found him work with circuses, as a professional bicycle racer and the like. He entered his first motorcycle race in 1908 astride an Indian he'd bought secondhand the previous year. His friends referred to him as 'Bake', the 'Cannonball' tag came later courtesy of a journalist writing about one of his epic rides. Baker entered a desert race in 1913 but retired with engine trouble, he set the hour/ distance record in 1914 as well as entering more desert races. In 1915 he made his famous transcontinental record run in a time of 11 and a half days. In 1916 he undertook The Three Flags Run with a new engine fitted to the bike he'd ridden across the continent on. This was another major feat of endurance, he left Vancouver, British Columbia on August 24th and arrived in Tijuana, Mexico on August 27th. He'd travelled 1655 miles in three days nine hours and 15 minutes having stopped only once to sleep – for three hours!

He went on to beat a number of motorcycle records in Australia and New Zealand and in 1919 attempted to beat the Sidecar Record for crossing the North American continent. Due to the atrocious state of the roads in the Midwest the bid failed. The motorcycle outfit was shipped back to the Indian factory very much the worse for wear with a broken frame caused by the poor condition of the roads. In 1922, aboard an Indian Scout, Baker made another transcontinental run travelling from New York City to Los Angeles in seven and a half days, a distance of 3,368 miles. Inevitably these records would fall as the roads were improved and bikes became faster but Baker stands out as a pioneer of early endurance record making, he never lost his enthusiasm for distance

A 1936 Sport Scout in race trim, in the pits at Daytona

rides. At the age of 60, in 1941, he made a long ride on an Indian 4. Cannonball Baker died in 1960.

A 1940 Sport Scout racer. The success of Sport Scouts would continue in the 1950s and 1960s in Class A hillclimbing, long after production had ceased

Adelen and Augusta Van Buren

Long distance riding wasn't an entirely male preserve as the Van Buren sisters proved in 1916. Adelen and Augusta rode their three speed Powerplus twins coast to coast during the course of a summer. They left Brooklyn, NY on the 4th of July and arrived in San Francisco on the 2nd September. They covered 3,300 miles and visited a number of Indian dealers on the way. A dealer in Omaha gave the girls a pistol for protection. Near Denver the pair made a detour to climb Pikes Peak, they were the first women solo riders to climb the 14,000 feet high mountain. They narrowly averted a disaster in Western Utah when they went 80 miles in the wrong direction along a trail, a prospector found them and gave them water. The Powerplus Indians performed faultlessly; one of the bikes suffered two punctures – the only reported problem.

Burton Albrecht

Burton Albrecht was an Indian factory sponsored rider and on the 31st of July 1931 recorded a 36 second mile at a California Speedway circuit. It was a record that would stand for the best part of twenty years.

Rody Rodenburg

In 1931 TT-type competition became popular in part due to Indian Advertising Executive Ted Hodgdon who attended such a race near New York. He was impressed by the fact that entrants could compete aboard their own ride-to-work machines. Subsequently he prevailed on the American Motorcycle Association to include this style of racing in the AMA rulebook. It was, and became the forerunner of Class C racing. Class C got a real boost when an AMA sanctioned 200 mile National Championship road race was organised in Jacksonville, Florida. The race, held on February 24th 1935 had something of a carnival atmosphere and included motorcycling celebrities in the role of officials. `Wild' Bill Cummins, an Indianapolis racer, was starter, Joe Petrali, a famous Harley rider was referee and the abovementioned `Cannonball' Baker was Chief Judge. The circuit was 1.6 miles long and 10,000 spectators lined the rails. An Indian and a Harley were neck and neck for the whole distance. The win went to Indian rider Rody Rodenburg who was on a Sport Scout. This win did nothing to harm sales!

Interest in the endurance and distance records with which much of this chapter is concerned lapsed somewhat as roads improved. The constant upgrading and construction meant that the distances between cities effectively shrank and that it was therefore no longer noteworthy to make such journeys on a motorcycle. One journey that was still deemed worthy of record making was Los Angeles to New York City. In 1934 the tarmac stretched 3,005 miles between the two cities. Randolph Whiting, a Chief rider, had two attempts at this run sequentially reducing the time taken. A time of four days and 19 hours was the faster of the two. His record fell to a Harley rider in the autumn of that year. Rody Rodenburg attempted the journey aboard a Sport Scout in the spring of 1935. He covered the 3,005 miles in two days 23 hours; to manage this massive reduction of the time he slept for only three hours.

Jeff Rosenburg's hybrid flat track Sport Scout, it's a 1939 motor but the alloy tank and the telescopic forks are later and probably of British origin

Ed Kretz

The name Ed Kretz is synonymous with Indian race wins. Kretz was born in San Diego in 1911 of Swiss-German parents and one of eleven children. During the Depression he earned a living driving trucks. He bought a secondhand VL Harley Davidson and went to watch a couple of races on it. For a particular race held at The Ascot Speedway he stripped the lights and mudguards off the VL and entered. The track consisted of a banked circuit of 0.625 miles and a route over an adjacent hill. Second place went to Kretz despite a puncture and being up against factory backed and professional riders. Floyd Clymer, then Indian dealer in LA and the race promoter, got Kretz a factory ride and employed him as a mechanic. Kretz didn't disappoint, in 1936 he won the National Championship 200 mile race in Savannah, Georgia. In 1937 the same event was moved to Daytona Beach, Kretz took the chequered flag again. The nickname 'Ironman' stuck and the races he won both sides of World War II became legend. Victories at Laconia, Daytona and Langhorne included a number of 100 and 200 mile Nationals. Kretz went on to run a successful Indian dealership in a suburb of Los Angeles.

Fred Ludlow

In 1937 a group of Indian fans under the direction of Hap Alzina took a number of machines out to the salt flats at Bonneville in Utah. For records to stand they have to be run over a measured distance two ways and the average time for both runs counts. Fred Ludlow took a Class C 45 cu in Sport Scout to an average of 115.226 mph to establish a new Class C record. On the same expedition Ludlow rode a 74 cu in Chief two ways to average 120.747 mph, another class record. The team also attempted a record in an enclosed streamlined machine known as The Arrow. Above certain speeds it got into a serious wobble and Alzina stopped its use stating that he was unprepared to risk anyone's life in it.

Floyd Emde

In 1948 Floyd Emde, aboard a prewar Sport Scout, took first place in the 200 mile Daytona race. To win, he fought off strong challenges from the 45 cu in WR Harley Davidsons and a large contingent of riders on imported British bikes.

In hill climbing weight reduction is everything, this bike has even had the cylinder fins ground down to save weight. The fuel tank is deliberately small for the same reason

Indians Today

As is to be expected amongst motorcyclists, there's still tremendous interest in Indian motorcycles. It is believed that in the USA the numbers of Indians registered for the road is actually increasing as more restored bikes are put back on the road! Racers, restorers, customisers and a few others still think highly of the Springfield machines. Among the latter group are some of the remaining wall of death riders. Once a common sight at fairs and carnivals around the world the `wall of death' still thrills. The characteristics of the Indian motorcycle that made it a favourite mount for those who earned a living on the vertical wooden walls in earlier years of the twentieth century mean it is still suitable for those short but dangerous circuits. The stable handling of the Indian 101 Scout which allowed the rider to let go of the handlebars, a lubrication system that didn't leave too much oil on the wooden wall and a cast T-manifold that the rider could pivot to keep the carburettor horizontal are three of those endearing characteristics.

Indian's large production and policy of exporting their machines from the earliest days mean that Indian enthusiasts are found in countries around the globe. Enthusiasts range from those who own a single machine to those who have massive collections spanning many of Indian's fifty two years in

Right
Alan Ford and Ned Kelly, two British Wall of Death riders who still use Indian motorcycles

Left
On the wall and thrilling the crowds ...

the business of motorcycle manufacture. When Indian were in their heyday dealers were encouraged to sponsor picnics and runs for their customers meaning that there's long been an active social scene associated with the motorcycles. As the dealers changed their franchises the social scene was refocussed on the various Indian owners and riders clubs as well as the new breed of dealers that dealt with Indian restoration parts.

One of the important events is Indian day which is held in the grounds of the Indian Motorcycle Museum which is based in part of the old Springfield factory. There's an annual European Indian rally and of course many Indians turn up at general historic motorcycle meets and races. Some of the military bikes appear at military vehicle rallies. Alongside all this activity are the autojumbles where restorers can find that elusive part to complete a rebuild, buy a complete bike in need of restoration or simply buy some Indian ephemera. New 'Old Stock' spares still in their original packets are both useful and nice to collect. Rarer items such as old racing jerseys fetch high prices as befitting reminders of days when Indian ruled the ovals.

Above
Indian Day at Springfield attracts the crowds and a huge variety of iron redskins

Left
Ken Young's Indian Chief attracting admirers!

An offshoot of the restoration scene are the customisers. Indians have long been modified to do a certain job better whether it was for racing or hillclimbing or simply 'bobbed' either side of World War Two to enhance street performance and style and this process continues today. It's possible to build or restore an Indian bobber or racer in order to own something that's both vintage and different. Taking this a step further there are a few Indian choppers about but the majority of customised Indians tend to be stock bikes with extra chroming and fancy paintwork. A favourite subject for the custom painters is paint relating to the native American peoples for obvious reasons.

Over the years there have been several attempts to resurrect the Indian company and produce bikes again. Many of these projects have come to nothing and others have had nothing more to do with Indian than the sticker on the tank – Taiwanese trial bikes for example. The most recent

A massive variety of parts, ranging from really rusty to brand new, is available and 'Sure it'll all come in useful someday'

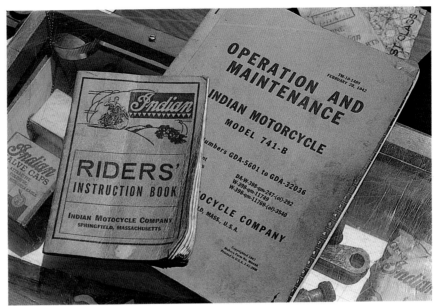

attempt to revive the Indian marque is being made by Wayne Baughman in Albuquerque, New Mexico. At the time of writing his company, Indian Motorcycle Manufacturing Incorporated, has issued press releases about the new Chief and photos of a prototype. Whether a new Chief will actually make it into production and onto the streets remains to be seen.

Above
Handbooks such as these, a civilian rider's manual and a military maintenance manual, are just one aspect of the collectible ephemera that goes along with Indians

Left
Be Leenes Indian bobber 'Burning Blue' is based on the bobbers of the forties and fifties but with addition of some more modern parts. It is a 1946 1200cc Chief in a standard frame but with Suzuki forks and disc brakes

Left

Tony Leenes who runs an Indian museum and restoration business built this bike for a German customer and based its styling on Wall of Death bikes but fitted lights and other parts required by law for road use

Above

Another bike built by Tony Leenes in his Lemmer, Holland premises this time for himself. He based it on a Class C racer. The special is based around a 45 cu in Sport Scout engine in an original frame

Snow Pearl is a chopper and was also
built by Tony Leenes, partially along
Harley lines with a softail frame and a
Mustang petrol tank

Above

Chris Ireland is well known for building show standard choppers and trikes in England. Just for fun he keeps this Indian 741-B rat chop around

Right

The sidevalve engine in the rat has acquired something of a venerable patina through age and use

Left

In recent years the big American bike events such as Daytona and Sturgis have the reputation of being Harley-Davidson rallies but the machines made by their valiant competitor, Indian, are most welcome as this Indian Chief on Sturgis Main Street shows

Above

This customised Chief and sidecar is typical of many custom Indians; airbrushed artwork, non-standard two-tone paintwork and extra chrome

Left

The glorious, enduring image of the Indian Chief, surely one of the most distinctive and immediately recognisable motorcycles ever produced

Overleaf

More of that image goes further into the heritage of the native American peoples, such as the rear fender on this Chief